SCIENCE WITH A BEAT

OUT BEYOND

By Jacquie Hawkins

CHAPTER ONE
Where was the sun all night? – P. 1

CHAPTER TWO
Can we fall off the Earth? – P. 5

CHAPTER THREE
What makes the colors in the rainbow? P. 9

CHAPTER FOUR
Why is the sky blue? – P. 13

CHAPTER FIVE
When are the days the longest? – P. 16

CHAPTER SIX
Why does smoke go up? – P. 19

CHAPTER SEVEN
Why does the moon change shapes? – P. 23

CHAPTER EIGHT
What are stars? – P. 26

CHAPTER NINE
Why does your shadow grow big and small? – P. 29

CHAPTER TEN
Why do the sun's rays give less heat in winter? P. 32

CHAPTER ELEVEN
Is air lighter on a mountain than in a valley? P. 36

CHAPTER ONE
WHERE WAS THE SUN ALL NIGHT?

The <u>sun</u> was right <u>there</u> in the
<u>very</u> same <u>spot</u>.
It is <u>you</u> who has <u>moved</u>, be<u>lieve</u> it or <u>not</u>.
<u>Our</u> big, round <u>earth</u> keeps on
<u>spinning</u> a<u>round.</u>
All <u>night</u> and all <u>day</u> on its <u>path</u> it is <u>bound</u>.

2.

When the <u>sun</u> comes <u>up</u> and
<u>moves</u> through the <u>sky</u>
It is <u>really</u> the <u>earth</u> taking <u>us</u> for a <u>ride</u>.
As it <u>moves</u> through the <u>sky</u> the earth
<u>takes</u> us <u>away</u>
Un<u>til</u> there's no <u>sun</u> at the <u>end</u> of the <u>day</u>.

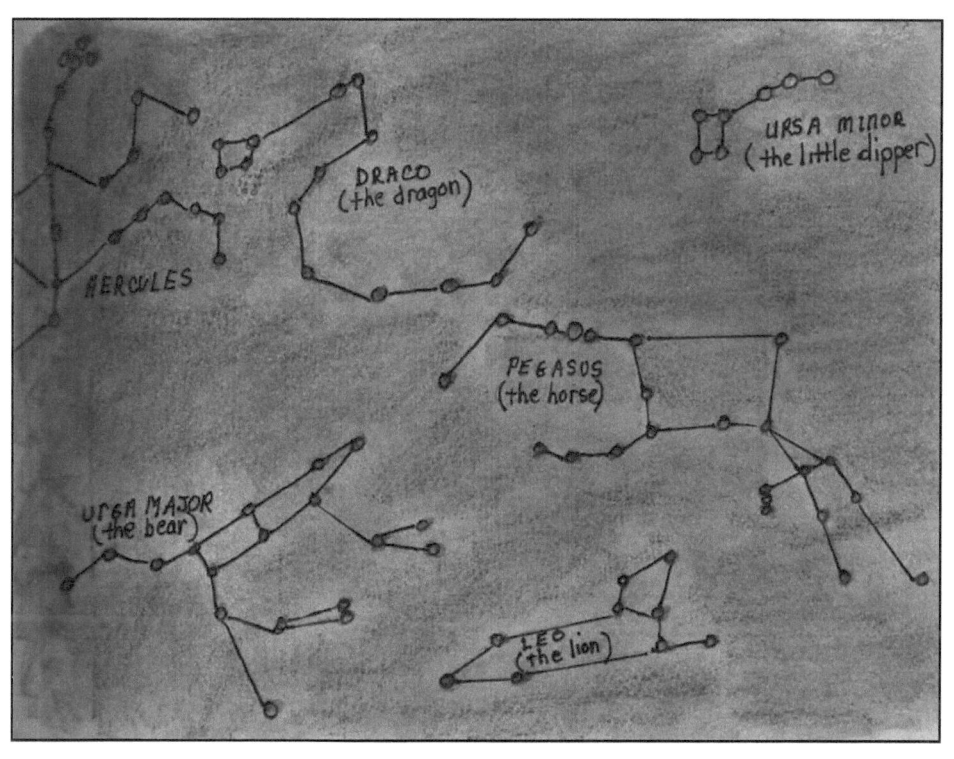

Be<u>cause</u> in the <u>night</u> we can<u>not</u> see the <u>sun,</u>
Stars <u>seem</u> to pop <u>out</u> in the
<u>sky</u> one by <u>one</u>.
But the <u>stars</u> were all <u>there</u> in the
<u>sky</u> all the <u>time</u>.
We <u>just</u> couldn't <u>see</u> them for the
<u>sun</u> made us <u>blind</u>.

4.

Somewhere in this world it
is dead of night
While we're having day and the
sunshine is bright.
And when we're asleep, safe and
warm in our beds,
There are children out playing in
sunshine instead.

CHAPTER TWO
CAN WE FALL OFF THE EARTH?

If the Earth is a sphere as it travels and spins
Why don't we fall off?
What keeps us all in?
And why, when you toss a ball
high in the air,
Does it come right back down?
It doesn't stay there!

And <u>why</u> when you <u>jump</u> do you <u>stop</u> way too <u>soon</u>
And <u>come</u> back to <u>Earth</u>?
Why not <u>jump</u> to the <u>moon</u>?
<u>I'd</u> like to <u>jump</u> up and <u>stay</u> in the <u>sky</u>
But <u>I</u> cannot <u>do</u> that.
I <u>want</u> to know <u>why</u>!

It's because of the Earth's pull

Don't you see?

This pull that's so strong

Is called gravity.

Gravity holds all the

Stars in their space

And keeps all the planets

Revolving in place.

8.

But you see, you and I do not have to fear
For gravity makes one thing very clear.
We'll never fall off this great planet of ours.
We're held into place with invisible bars.

CHAPTER THREE
WHAT MAKES THE COLORS IN THE RAINBOW?

The sunlight you see may look yellow to you
Or maybe to you it looks white.
But really it's violet, yellow and blue,
Green, orange and red light.

10.

When all of these colors are
mixed up together
A white light is all that you see.
But when the sunlight streams
down through a raindrop
It acts as a prism. You see?

A <u>raindrop</u>, or <u>prism</u>, di<u>vide</u>s up the <u>light</u>.
Like <u>magic</u> it <u>pulls</u> out the <u>colors</u>
<u>That</u> you and <u>I</u> cannot <u>normally</u> <u>see</u>
Be<u>cause</u> they're all <u>jumbled</u> <u>together</u>.

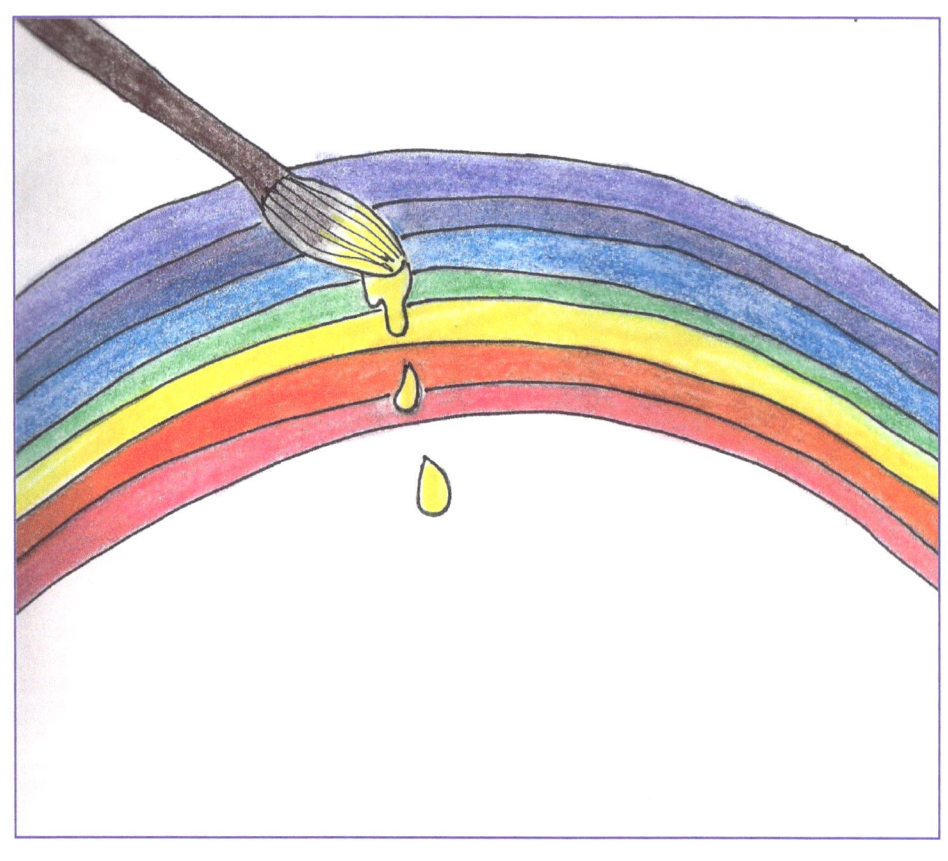

And so a rainbow we see in the sky
In a beautiful, colorful band
Of red, orange yellow green, blue, and, violet
As if by an artist's hand.

CHAPTER FOUR
WHY IS THE SKY BLUE?

What we call sky is nothing but air
With billions of dust specks floating up there
Along with the sun's light of violet and blue,
Green, yellow, orange, plus red rays too.

14.

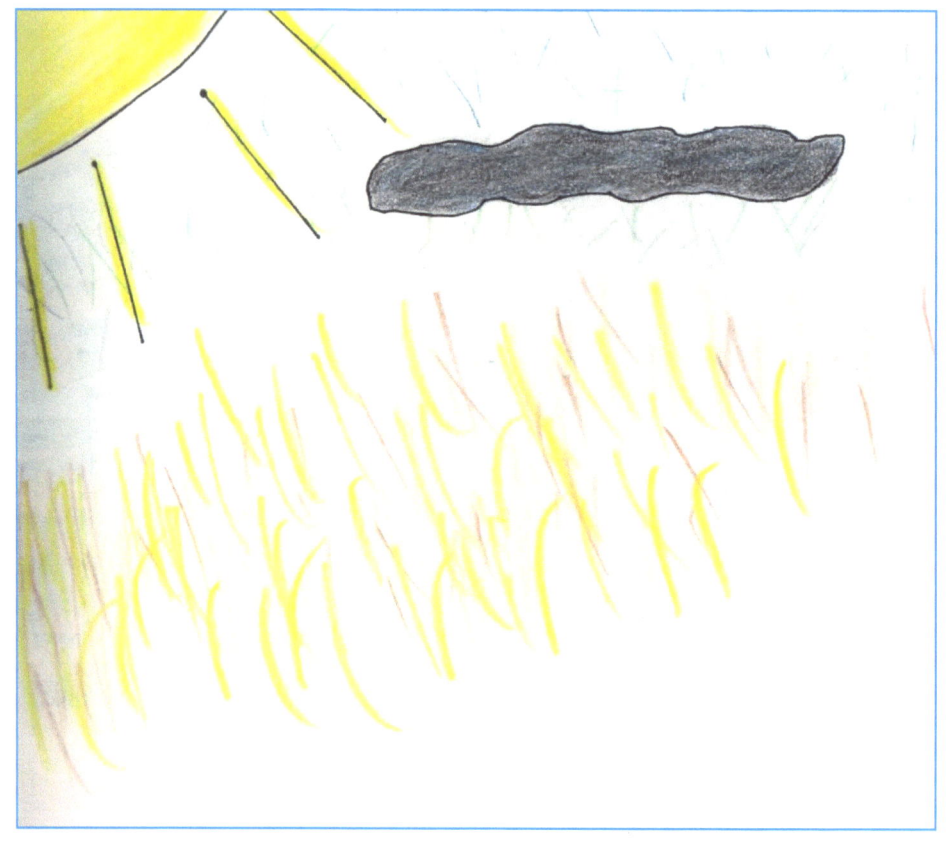

You won't see these colors,
although you can try,
Unless there's a rainbow that's
stretched through the sky.
The yellow and orange rays
come straight to earth.
The violet-blue-green rays are
scattered out first.

The <u>dust</u> in the <u>air</u> makes them
<u>bounce</u> all <u>around</u>
<u>Staying</u> up <u>high</u> before
<u>coming</u> back <u>down</u>.
<u>When</u> we look <u>up</u> we see <u>all</u> the blue <u>rays</u>
<u>Reflected</u> by dust in <u>a beautiful way</u>.

CHAPTER FIVE
WHEN ARE THE DAYS THE LONGEST?

When the sun shines brightly in your
room in the morning
And doesn't want you in your bed to stay;
When it seems to fill up your
room with soft sunbeams that
Beckon you to come outside and play;

When you <u>don't</u> need a <u>jacket</u> or even a <u>sweater</u>;
When <u>you</u> shed your <u>socks,</u> going <u>with</u> naked <u>feet</u>;
When <u>you</u> swim and <u>boat</u> or just <u>play</u> in the <u>sprinkler</u>
And <u>love</u> lots of cold <u>things</u> to <u>drink</u> and to <u>eat</u>;

18.

That's when the days are the very longest
And the darkness of nighttime is very short.
It's like the sun knows that
kids like the summer
And is trying to be a really good sport.

CHAPTER SIX
WHY DOES SMOKE GO UP?

The gray-black smoke from a blazing fire
Is made from the littlest specks of ash
That the fire does not seem to
want to burn-up
And so all this junk it just tries to trash.

20.

All of the air just above the hot fire
Is not only hot….it is VERY hot.
But all of the air is very much colder
That you will find just above that spot.

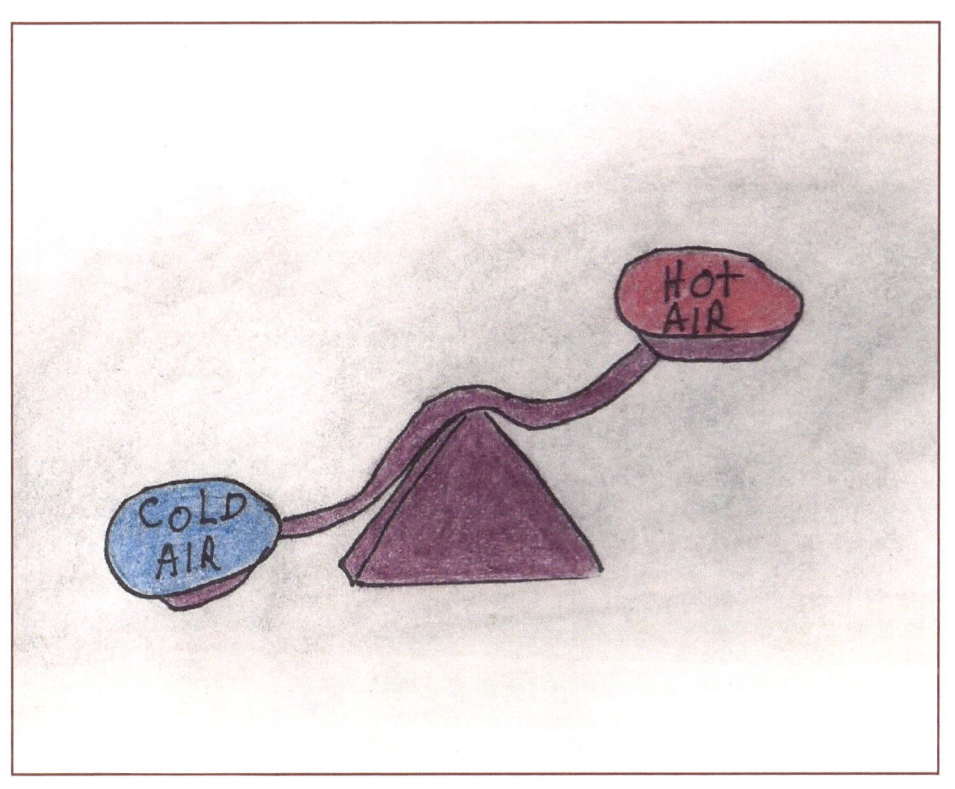

That really cold air is certainly heavy…
Too heavy for all of it to be held-up…
And so it all falls back down into the fire
Pushing back up all that smoky smut.

22.

The hot smoky air is so very light that it
Drifts its way up and goes out into space.
It will keep drifting and floating on upwards
Leaving behind nothing…no, not a trace.

CHAPTER SEVEN
WHY DOES THE MOON CHANGE SHAPE?

Just like the earth, the moon
gives off no light.
Our big sun shines on it and
makes it look bright.
So the half of the moon that the
sun hits will shine
And the other half of it is dark all the time.

24.

At <u>times</u> a thin <u>sliver</u> is <u>all</u> we can <u>see</u>
For the <u>half</u> that is <u>lighted</u> we <u>just</u> see <u>partly</u>.
As the <u>moon</u> con<u>tinues</u> to <u>circle</u> the <u>world</u>
It may <u>look</u> like a <u>yellow</u> ball
<u>some</u>body <u>hurled</u>.

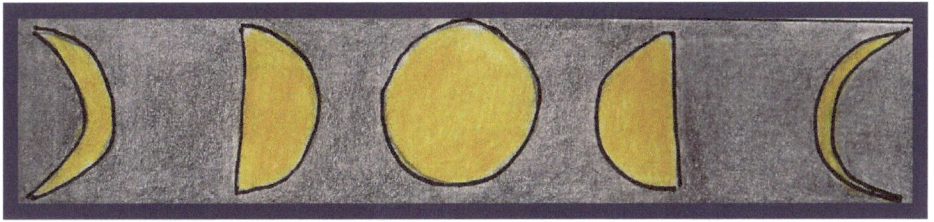

There is <u>one</u> time each <u>month</u> that we <u>call</u> the New <u>Moon</u>,
When the <u>moon</u> can't be <u>seen</u>.
and <u>yet</u> very <u>soon</u>
It will <u>grow</u> to a <u>sliver</u>, then <u>back</u> to a <u>ball</u>
And at <u>times</u> it will <u>shrink</u> back to <u>nothing</u> at <u>all</u>!

CHAPTER EIGHT
WHAT ARE STARS?

The <u>stars</u> are great <u>balls</u> of real
<u>hot</u>, glowing <u>gases</u>
<u>Whirling</u> and <u>twirling</u> in <u>space</u>
in huge <u>masses</u>.
Our <u>sun</u> is a <u>star</u>, not a <u>large</u> one in <u>fact</u>.
<u>Most</u> are <u>bigger</u>. <u>Just</u> think of <u>that</u>!

The stars look so tiny when
they're up so high,
While our sun looks huge right
there in the sky.
But those tiny pinpoints of light that you see
May be billions of miles from
you and from me.
That is the reason that they look so small.
Aren't stars just the prettiest
sight of them all?

All <u>night</u> they just <u>twinkle</u> and
<u>show</u> off their <u>light</u>…
Some <u>blue</u>; some are <u>red</u>,
some <u>yellow</u> or w<u>h</u>ite.
I <u>wonder</u> if <u>fairies</u> who <u>lived</u> far <u>beyond</u>
<u>Scattered</u> them <u>there</u> with a
<u>twist</u> of their <u>wands</u>?

29.
CHAPTER NINE
WHY DOES YOUR SHADOW GROW BIG AND SMALL?

That light cannot shine through a person's a fact.
Your body will block light if you're thin or fat.
The dark spot in front's where
the sun cannot come.
The length of your shadow's
because of the sun.

At <u>noontime</u> in <u>summer</u> light
<u>rays</u> comes straight <u>down</u>
Hitting <u>your</u> head and <u>shoulders,</u> then
<u>straight</u> to the <u>ground</u>.
Your <u>body</u> won't <u>block</u> much.
Your <u>shadow</u> is <u>neat</u>,
Tucked <u>safely</u> and <u>out</u> of sight
<u>beneath</u> your <u>feet</u>.

But <u>when</u> the sun <u>rises</u> or <u>sets</u> in the <u>west</u>
The <u>light</u> hits your <u>head,</u> shoulders
<u>plus</u> all of the <u>rest</u>.
Your <u>body</u> blocks <u>light</u> and makes
<u>your</u> shadow <u>stretch</u>.
It <u>grows</u> tall and <u>thin</u> making <u>you</u> laugh, I <u>bet</u>.

CHAPTER TEN
WHY DO THE SUN'S RAYS GIVE LESS HEAT IN THE WINTER?

In <u>winter</u> the <u>sun</u> seems so <u>bright</u> in the <u>sky</u>
But it <u>won't</u> warm you <u>up</u> as much.
<u>Do</u> you know <u>why</u>?
As <u>you</u> watch the <u>winter</u> sun
<u>climb</u> it seems <u>shy</u>
Be<u>cause</u> it will <u>not</u> make it <u>up</u> quite so <u>high</u>.

At <u>noon</u>time, in <u>summer</u>,
the <u>sun</u> is <u>above</u>
But in <u>winter</u> it's <u>not</u> as high <u>as</u> it once <u>was</u>.
<u>Instead</u> it's much <u>lower</u> though <u>it</u> really <u>tries</u>.
The rays <u>strike</u> from a <u>slant</u> and get
<u>into</u> your <u>eyes</u>.

34.

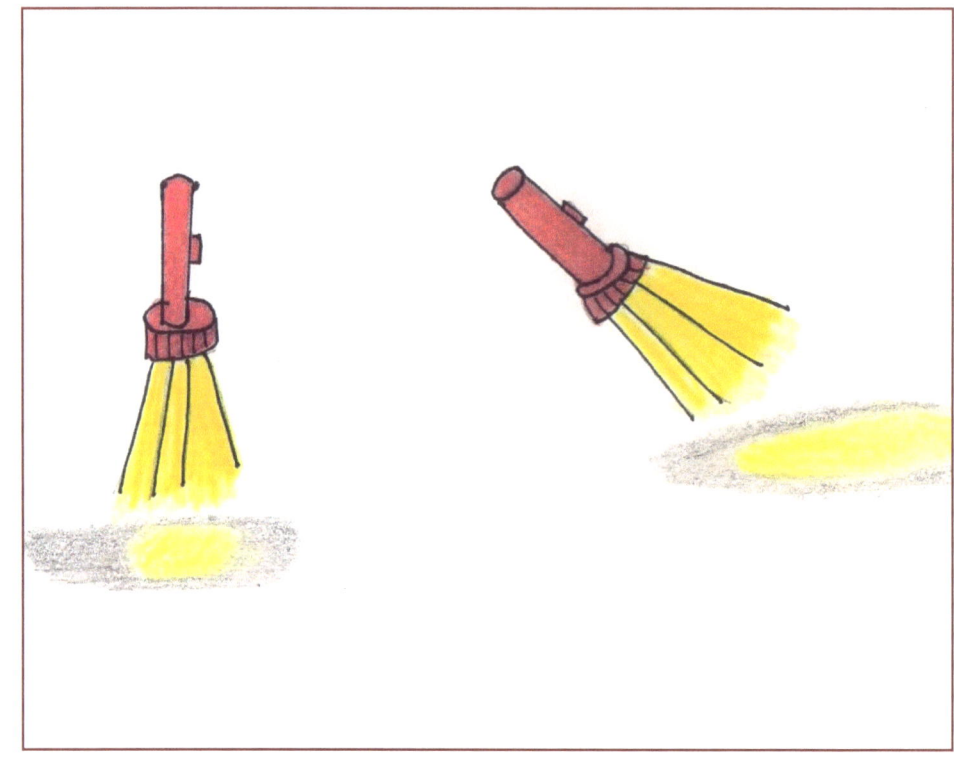

Pick up a flashlight that's lying around.
Pretend it's the sun and
shine it straight down.
You'll see that the spot that it
makes looks quite round.
Then slant it. The light covers
up much more ground.

In <u>winter</u> the <u>sunlight</u> hits <u>earth</u> at a <u>slant</u>.
We <u>may</u> wish it <u>wouldn't</u> but to
<u>change,</u> well, it <u>can't</u>.
So the <u>sun's</u> slanted <u>rays</u> spread
<u>out</u> all the <u>heat</u>
And the <u>job</u> that it <u>does</u> seems to
<u>be</u> incom<u>plete</u>!

CHAPTER ELEVEN
IS AIR LIGHTER ON A MOUNTAIN THAN IN A VALLEY?

Up in the mountains and high in the sky
The air is much thinner but
do you know why?
When one climbs a mountain it's
hard to get there
Because there's less oxygen high in the air.

Where <u>pilots</u> in space crafts fly
<u>up</u> really <u>high,</u>
Up <u>there</u> where we <u>can't</u> even
<u>see</u> them fly <u>by</u>,
<u>There</u> is no <u>air</u> and they <u>would</u> surely <u>gasp</u>
And <u>die</u> if not <u>wearing</u> an <u>oxygen</u> <u>mask</u>.

38.

If a <u>cork</u> you push <u>under</u> the <u>water</u> one <u>night</u>
It will <u>pop</u> back up <u>quickly</u>
because it is <u>light</u>.
A <u>balloon</u> filled with <u>helium</u>,
<u>because</u> it's so <u>light,</u>
Will <u>seem</u> to float <u>up</u> and sail
<u>quite</u> out of <u>sight</u>.

But it <u>won't</u> go <u>forever</u>, wher<u>ever</u> it's <u>bound</u>,
For it <u>will</u> be as <u>light</u> as the <u>air</u> that's <u>around</u>.
As the <u>air</u> thins <u>out</u> it <u>eventually</u> goes <u>down</u>
Un<u>til</u> once <u>again</u> it will <u>land</u> on the <u>ground</u>.

Down in a valley the air is quite heavy
While up on a mountain it gets very thin
So people who live in the
Andes and Rockies
Don't always have too much vigor and vim.

LOOK FOR THESE OTHER BOOKS FROM THE SCIENCE WITH A BEAT SERIES:

OUR WONDERFUL PLANET EARTH

WEATHER

WATER WORLD

www.ingramcontent.com/pod-product-compliance
Lightning Source LLC
Chambersburg PA
CBHW040929180526
45159CB00002BA/665